$$W_{end} = \left(\left(\frac{(3n+1)P_s}{73(days)^2}\right) + S_d\right) \text{ years}$$

$$R_{mes} = 1(T_{sun})(20 \text{ years})$$

$$R_{mes} = Y * 1day$$

$$R_{mes} = 12.360065 \ R_{me}$$

$$W_{end} = X + (S_d) \text{ years}$$

$$M_{Fin} = \left(\left(\frac{20P_s}{1dia}\right) + F_{ini}\right) \text{ años}$$

The Formula Bárcena

$$R_{Les} = 12.360065 \ R_{Le}$$

$$R_{Les} = Y * 1dia$$

$$W_{end} = \left(\left(\frac{20P_s}{1day}\right) + S_d\right) \text{ years}$$

$$M_{Fin} = X + (F_{ini}) \text{ años}$$

$$R_{Les} = 1(V_{Sol})(20 \text{ años})$$

$$W_{end} = \left(\left(\frac{R_{me}}{365(days)^2}\right) + S_d\right) \text{ years}$$

$$M_{Fin} = \left(\left(\frac{(3n+b)P_s}{73(dias)^2}\right) + F_{ini}\right) \text{ años}$$

$$M_{Fin} = \left(\left(\frac{R_{Le}}{365(dias)^2}\right) + F_{ini}\right) \text{ años}$$

1

The Formula Bárcena

By Jesús E. Bárcena

2015

The formula Bárcena

Copyright © 2015 by Jesús E. Bárcena

Library of Congress Cataloging-in-publication Data

Bárcena, Jesús. The formula Bárcena

This book is a work of not-fiction.

Acknowledgements

I would like to thank the staff at the Editorial department.

This book is dedicated to god, my family and daughters Richelle and Kimberly

CENTURIA I

XLVIII Bing ans du regne de la Lune passez,

Sept mille ans autre tiendra sa monarchie :

Quand le Soleil prendra ses iours lassez,

Lors accomplir & mine ma prophetie.

XLVIII Twenty years after the moon kingdom,

Seven thousand years another will have its monarchy :

When the sun take their weary days,

Then my prophecy fulfilled and consumed.

P_s is a variable that changes between two types of constant values if you consider the sideral-Orbital period or synodic period, but if you consider reducing the constans receives two values for each of them.

CENTURIA I: XLVIII

Twenty-year reign of the moon is a simple formula and is as follows:

- $1^P{}_s$ is a cycle of 20 year reign.

- The years convert into days.

- Plus the number of years made the revelation to Michel de Nostradamus.

A few years before the end of the Sun will lose energy and the constant 29.530589 will change to 29.16666666666666666666666666669 and the earth will have another satellite 7000 years.

The question that the world is doing now is since when has counting the 7th millennium?

The answer is to that question is since they made the revelation to Michel Nostradamus.

In this book will demonstrate that the years obtained with the formula Bárcena and 7000 years are equivalent.

In that year the sun will shut off, this event is called to the weary days and it will generate the change of position of the moon with another satellite and when this phenomenon occurs this planet will be dead.

Graphic of the reign of the moon and the value of the Synodic period

CENTURIA I : XLVIII

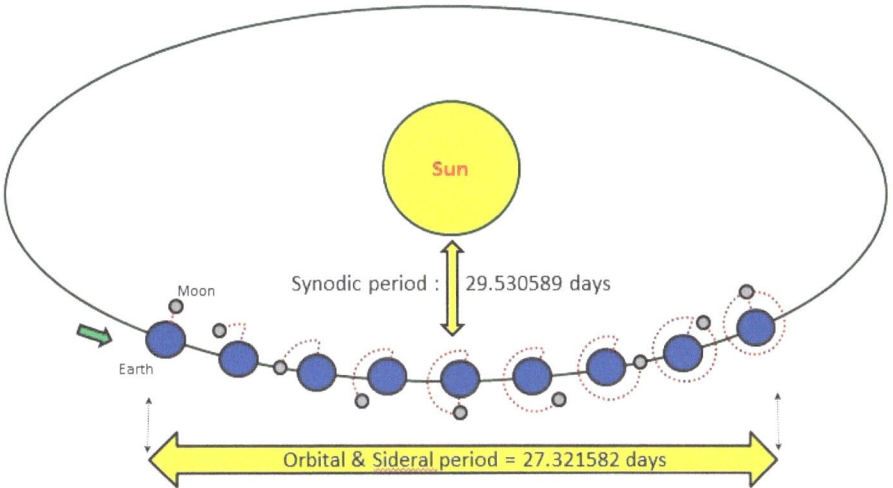

Sun

Moon

Synodic period : 29.530589 days

Earth

Orbital & Sideral period = 27.321582 days

A-1

Explaining the chart A-1 completely the Sun in the center and the earth around the Sun and showing the path that makes the moon around the earth.

The cycle of the moon is observed from earth 27.321582 days and is called this cycle and Sideral and Orbital period.

The same cycle of the moon is 29.530589 days observed from space and that cycle is called synodic period.

In this book we will call that reign cycle of the moon.

When you add up the value of X at Sd you will have the date of the end of the world.

$$W_{end} = X + (^{S}d) \text{ years}$$

CENTURIA I: XLVIII

Explaining this centuria in a different way with the following formula:

$$W_{end} = X + (^{S}d) \text{ years}$$

$$W_{end} = ((\frac{R_{me}}{365(days)}) + ^{S}d) \text{ years}$$

$$W_{end} = (20^{P}s + ^{S}d) \text{ years}$$

$$P_{s} = \text{(Orbital and Sideral) period}$$

$$P_{s} = 27.321582 \text{ days}$$

The following P_{s} is another option:

26.90464908124045701047754435845,
26.923076923076923076923076923077.

T_{sun} = (Number of circle around of the sun)

T_{sun} = (Number of P_s)

T_{sun} = 13.35940210197P_s

T_{sun} = 365 days

P_s = Synodic period

P_s = 29.530589 days

The following P_s is another option:

29.14670317134382842801733972166,
29.16666666666666666666666666666669

$T_{sun} = (^{\#}P_s)$

T_{sun} = 12.36006501597 P_s

T_{sun} = 365 days

S_d = Start date

S_d = Number of the year that Michell knows what will happen in this event

S_d = (1537–1555)

n = 1(Normal years) = 365 days

l = 1(Leap years) = 366 days

R_{me} = X

R_{me} = 1(Reigned in the land)

$R_{me} = 1(P_s)((15\,n) + (5\,l))$

$R_{me} = 1(P_s)(20\text{ years})$

R_{mes} = Y

R_{mes} = 1(Reigned in the Sun)

R_{mes} = 1(T_{sun}) (20 years)

R_{mes} = 1(T_{sun}) ((15n) + (5l))

If you convert decimals (R_{me}) will get the month, day, time and exact second.

W_{end} = Sun takes his days tired

W_{end} = End of the World

W_{end} = X + (S_d) years

X = It is a number of days and belongs to a part of the formula Bárcena.

This X is converted to years and add up the (X + S_d) we get the day that the Sun takes her tired days.

If they consider the Synodic period the value of P_s changes to:

P_s = 29.530589 days

P_s = 1(Reigned in the land)

P_s = (Synodic) period

Both values are then derived from the current and original constant 29.530589 these two values (a and b) days are obtained from the constant of the synodic period which replaced the formula R_{Les} will obtain the parameter of 7000 years.

a.- 29.14670317134382842801733972166

b.- 29.166666666666666666666666666669

R_{me} = X

$$R_{me} = 1(^P s)\ (20\ \text{years})$$

$$R_{me} = 1(^P s)\ ((15\ ^n) + (5\ ^l))$$

$$R_{mes} = Y$$

$$R_{mes} = 1(^T sun)\ (20\ \text{years})$$

$$R_{mes} = 1(^T sun)\ ((15\ ^n) + (5\ ^l))$$

If convert decimals of R_{me} and subtracts $(X - {}^C F)$ you will help get the month, day, time and exact second when had the vision to Michel de Nostradamus and knew that the Sun will take his weary days.

W_{end} = The Sun takes her tired days, end of the world.

$$W_{end} = X + (^S d)\ \text{years}$$

$$W_{end} = (\, (\frac{R_{me}}{365(days)}) + S_d) \text{ years}$$

$$W_{end} = (20P_s + S_d) \text{ years}$$

C_F = May 20

C_F = (Date of the Centuria IX - LXXXIII)

S_d = Subtraction of X and C_F

$S_d = X - C_F$

Developing the formula with the Orbital period 27.321582 and considering leap year

$$W_{end} = \left(\left(\frac{(3n + l)P_s}{73(days)^2} \right) + S_d\right)$$

years

CENTURIA I: XLVIII

Explaining this Centuria in a different way with the following formula:

$$R_{me} = 1(^Ps)(20 \text{ years})$$

The question is when? And the answer is:

$$X = R_{me}$$

$$R_{me} = 1(^Ps)((15\,n) + (5\,l))$$

$$R_{me} = (^Ps)((15*365) + (5*366)) \text{ days}$$

$$R_{me} = 1(^Ps)\,(5475+1830) \text{ days}$$

$$R_{me} = (27.321582*7305) \text{ days}$$

$$X = 199584.15651 \text{ days}$$

Converted days into years:

X = 199584.15651 (days) (1 year/365 days)

X = 546.805908 years

Converted year into days: 365 days divide between 12 months = 30.416667 days;

AVERAGE = 30 days

Converted years into days:

0.805908 (years) (365 days/1 year)

= 294.15642 days

Converted days into months:

294.15642 (days) (1 month/30 days)

= 9.805214 months

Converted months into days:

0.805214 (months) (30 days/ 1 month)

= **24**.15642 days

Converted days into hours:

0.15642 (days) (24 hours/1 day)

= **3**.75408 hours

Converted hours into minutes:

0.75408 (hours) (60 minutes/1 hour)

= **45**.2448 minutes

Converted minutes into seconds:

0.2448 (~~minutes~~) (60 seconds/ 1 ~~minute~~)

= 14.688 seconds

Approximately: X = 546 years, 9 months, 24 days, 3 hours,   45 minutes, 14 seconds

Recalling the following:

C_F = May 20

C_F = (Date of the centuria IX - LXXXIII)

S_d= Subtraction of X y C_F

$S_{d=}$ X $- C_F$

S_d = (1537-55) years, 7 months, (26 ó 15) days, 20 hours, 14 minutes, 46 seconds

Recalling that:

X = R_{me}

We also know that:

$$R_{me} = 20P_s \text{ years}$$

$$R_{me} = (\overline{\frac{(3n + l)P_s}{73(days)}}) \text{ years}$$

$$W_{end} = ((\overline{\frac{(3n + l)P_s}{73(days)}}) \text{ years} + S_d)$$

Replacing the values of each variable and obtain the exact date of the end of the world.

$$W_{end} = (X + S_d)$$

Saturday, May 20, 2084

Thursday, May 11, 2084

Developing the formula with Orbital 27.321582 period and without considering leap year

$$W_{end} = \left(\frac{20P_s}{1day} + S_d \right) \text{ years}$$

CENTURIA I: XLVIII

Explaining this centuria in a different way with the following formula:

$$R_{me} = 1(^P s)(20 \text{ years})$$

$$R_{me} = (^P s)(20*365) \text{ days}$$

$$R_{me} = 1(^P s)(7300) \text{ days}$$

$$R_{me} = (27.321582*7300) \text{ days}$$

X = 199447.5486 days

Converted days into years:

X = 199447.5486 (days) (1 year/365 days)

X = 546.43164 years

The years convert into months: 365 days divide between 12 months = 30.416667 days; AVERAGE = 30 days

Converted years into days:

0.431640000000016 (years) (365 days/1 year)

= 157.548600000006 days

Converted days into months:

157.548600000006 (days) (1 month/30 days)

= 5.25162000000019 months

Converted months into days:

0.25162000000019 (months) (30 days/ 1 month) = 7.5486000000057 days

Converted days into hours:

0.548600000005701 (~~days~~) (24 hours/1 ~~day~~)

= 13.1664000001368 hours

Converted hours into minutes:

0.166400000136818 (~~hours~~) (60 minutes/1 ~~hour~~)
= 9.98400000820908 minutes

Converted minutes into seconds:

0.98400000820908 (~~minutes~~) (60 seconds/ 1 ~~minute~~)

= 59.0400004925448 seconds

Approximately: X = 546 years, 5 months, 7 days, 13 hours, 9 minutes, 59 seconds

Recalling the following:

$$C_F = \text{May 20}$$

$$C_F = (\text{Date of the Centuria IX - LXXXIII})$$

$$S_d = \text{Subtraction of X and } C_F$$

$$S_d = X - C_F$$

$$S_d = (1537 - 55) \text{ years, } (0 \text{ ó } 12) \text{ months, } (12 \text{ ó } 3) \text{ days, } 10 \text{ hours, } 50 \text{ minutes, } \quad 1 \text{ seconds}$$

Recalling that:

$$X = R_{me}$$

We also know that:

$$R_{me} = 20P_s \text{ years}$$

$$W_{end} = (20^{P_s} \text{ years} + {}^{S_d})$$

Replacing the values of each variable and obtain the exact date of the end of the world.

$$W_{end} = (X + {}^{S_d})$$

Thursday, May 20, 2083

Tuesday, May 11, 2083

Developing the formula with the Orbital period 26.90465 and considering leap year

$$W_{end} = ((\frac{(3n + l)P_s}{73(days)^2}) + S_d)$$

years

CENTURIA I: XLVIII

Explaining this centuria in a different way with the following formula:

$$R_{me} = 1(^{P}s)(20 \text{ years})$$

The question is when? And the answer is:

$$X = R_{me}$$

$$R_{me} = 1(^{P}s)((15\,^{n}) + (5\,^{l}))$$

$$R_{me} = 1(^{P}s)((15*365) + (5*366)) \text{ days}$$

$$R_{me} = 1(^{P}s)(5475+1830) \text{ days}$$

$$R_{me} = \underline{(26.90464908124045701047754435845 * 7305)}$$
$$\underline{\text{days}}$$

X = 196538.46153846153846153846153845 days

Converted days into years:

X = 196538.46153846153846153846153845
(days) (1 year/365 days)

X = 538.46153846153846153846153846158 years

The years convert into days: 365 days divide
between 12 months = 30.416667 days;
AVERAGE = 30 days

Converted years into days:

0.46153846153846153846153846158
(years) (365 days/1 year)

= 168.46153846153846153846153846475 (days)

Converted days into months:

168.46153846153846153846153844475
(days) (1 month /30 days)

= 5.6153846153846153846153846614916 months

Converted months into days:

0.6153846153846153846153846614916
(months) (30 days/1 month)

= 18.46153846153846153846153844748 days

Converted days into hours:

0.46153846153846153846153844748
(days) (24 hours/1 day)

= 11.076923076923076923076922273952 hours

Converted hours into minutes:

0.0769230769230769230769273952
(hours) (60 minutes/1 hour)

= 4.615384615384615384153643712 minutes

Converted minutes into seconds:

0.615384615384615384153643712
(minutes) (60 seconds/ 1 minute)

= 36.92307692307692307921862272 seconds

Approximately: X = 538 years, 5 months, 18 days, 11 hours, 4 minutes, 36 seconds

Recalling the following:

C_F = May 20

C_F = (Date of the Centuria IX - LXXXIII)

S_d = Subtraction of X and C_F

$S_d = X - C_F$

S_d = (1537 – 55) years, (0 ó 11) month, (1 ó 23) days, 12 hours, 55 minutes,   24 seconds

Recalling that:

$X = R_{me}$

We also know that:

$$R_{me} = 20P_s \text{ years}$$

$$R_{me} = \left(\frac{(3n + l)P_s}{73(days)}\right) \text{ years}$$

$$W_{end} = \left(\left(\frac{(3n + l)P_s}{73(days)}\right) \text{ years} + S_d\right)$$

Replacing the values of each variables gets the exact date of the end of the world.

$$W_{end} = (X + S_d)$$

Monday, May 20, 2075

Saturday, May 11, 2075

Developing the formula with Synodic 26.90465 Orbital and without considering leap year

$$W_{end} = (\frac{20P_s}{1day} + S_d) \text{ years}$$

CENTURIA I: XLVIII

Explaining this centuria in a different way with the following formula:

$$R_{me} = 1(^P s)(20 \text{ years})$$

The question is when? And the answer is:

$$X = R_{me}$$

$$R_{me} = 1(^P s)(20 \text{ years})$$

$$R_{me} = 1(^P s)(20*365) \text{ days}$$

$$R_{me} = 1(^P s) (7300) \text{ days}$$

$$R_{me} = \underline{(26.92307692307692307692307692307077*7300)} \text{ days}$$

X = 196538.461538461538461538461538 4615384 days

Converted days into years:

X = 196538.461538461538461538461538461538 5384
(days) (1 year/365 days)

= 538.46153846153846153846153846154615 years

Converted days into years:

365 days divide between 12 months

= 30.416667 days; AVERAGE = 30 days

Converted years into days:

0.461538461538461538461538461538461538 4615 (years)
(365 days/1 year)

= 168.461538461538461538461538461538475 (days)

Converted days into months:

<u>168.461538461538461538461538 4475</u>
<u>(days) (1 month /30 days)</u>

= 5.615384615384615384615384614916 months

Converted months into days:

<u>0.615384615384615384615384614916</u>
<u>(months) (30 days/1 month)</u>

= 18.461538461538461538461538844748 days

Converted days into hours:

<u>0.461538461538461538461538844748</u>
<u>(days) (24 hours/1 day)</u>

= 11.076923076923076923076922273952 hours

Converted hours into minutes:

0.0769230769230769230769230769273952
(hours) (60 minutes/1 hour)

= 4.615384615384615384615643712 minutes

Converted minutes into seconds:

0.615384615384615384615643712
(minutes) (60 seconds/ 1 minute)

= 36.923076923076923076921862272 seconds

Approximately: X = 538 years, 5 months, 18 days, 11 hours, 4 minutes, 36 seconds

Recalling the following:

C_F = May 20

C_F = (Date of the Centuria IX - LXXXIII)

S_d = Subtraction of X and C_F

S_d = X − C_F

S_d = (1537 − 55) years, (0 ó 11) months, (1 ó 22) days, 12 hours, 55 minutes,   24 seconds

Recalling that:

X = R_{me}

We also know that:

$$R_{me} = 20P_s \text{ years}$$

$$W_{end} = (20P_s \text{ years} + S_d)$$

Replacing the values of each variables gets the exact date of the end of the world.

$$W_{end} = (X + S_d)$$

Monday, May 20, 2075

Saturday, May 11, 2075

CENTURIA I: XLVIII

Explaining this centuria in a different way with the following formula:

$$R_{mes} = 1(T_{sun}) \text{ (20 years)}$$

If you consider Synodic or Orbital period don't change the value to T_{sun}

$$(T_{sun} = 13.35940210197 P_s)$$

$$R_{mes} = Y$$

$$R_{mes} = (T_{sun}) ((15n) + (5l))$$

$$R_{mes} = (13.35940210197 P_s)((15*365) + (5*366)) \text{days}$$

$$R_{mes} = 13.35940210197 R_{me}$$

Y = (13.35940210197)(27.321582*7305)days

Y = 2666324.9999996034593247 days

The days convert into years:

Y = 2666324.9999996034593247 (days)(1 year/365 days)

Y = 7304.9999999989135871909589041 years

If you consider only two decimals you will has:

Y = 7305

If you consider only 13 without decimals you will has:

Y = (13) (27.321582*7305) days

Y = 2594594.03463 days

The days convert into years:

Y = 2594594.03463 (days) (1 year/365 days)

Y = 7108.47680720547945205479452054 7

If you consider only two decimals you will has:

Y = 7108

L

CENTURIA I: XLVIII

Explaining this centuria in a different way with the following formula:

If you consider Synodic or Orbital period don't change the value to T_{sun}

$(T_{sun} = 13.35940210197 P_s)$

$R_{mes} = 1(T_{sun}) \text{ (20 years)}$

$R_{mes} = Y$

If you consider only two decimal you will has:

$R_{mes} = (T_{sun}) \text{ (20 years)}$

$R_{mes} = (13.35940210197 {}^{P}{}_{s})\ (20\ years)$

$R_{mes} = 13.35940210197\ {}^{R}{}_{me}$

Y = (13.35940210197) (27.321582*7300) days

Y = 2664499.999999603730742 days

The days convert into years:

Y = 2664499.999999603730742 (days) (1 year/365 days)

Y = 7299.9999999989143308 years

If you consider only two decimals you will has:

Y = 7300

If you consider only 13 without decimals you will has:

Y = (13) (27.321582*7300) days

Y = 2592818.1318 days

Y = 2592818.1318 (days) (1 year/365 days)

Y = 7103.61132

Y = 7104

CENTURIA I: XLVIII

Explaining this centuria in a different way with the following formula:

$$R_{mes} = 1(^T sun) \, ((15 \, ^n) + (5 \, ^l))$$

$$R_{mes} = Y$$

If you consider Synodic or Orbital period the value P_s change to:

$$P_s = 26.9046490812404570104775445845 \text{ days}$$

Recalling the following:

T_{sun} = (Number of turns in the Sun)

$^P s$ = 13.35940210197 $^P s$

If you consider Synodic or Orbital period don't change the value to T_{sun}

(T_{sun} = 13.35940210197$^P s$)

R_{mes} = 1(T_{sun}) (20 years)

R_{mes} = Y

R_{mes} = (T_{sun}) $\underline{((15^n) + (5^l))}$

R_{mes} = (13.359402) $\underline{(^P s) ((15*365) + (5*366))}$ days

R_{mes} = 13.359402$^{R}{me}$

If you consider only 13 without decimals you will has:

Y = (13) (26.9046490812404570104775443584S*7305) days

Y = 2555000 days

The days convert into years:

Y = 2555000 (days) (1 year/365 days)

Y = 7000 years

O

CENTURIA I: XLVIII

Explaining this centuria in a different way with the following formula:

$$R_{mes} = 1(^{T}sun) \text{ (20 years)}$$

$$R_{mes} = 1(^{T}sun) \text{ (20*365) days}$$

$$R_{mes} = Y$$

If you consider Synodic or Orbital period the value ^{P}s change to:

$$^{P}s = 26.923076923076923076923076923077 \text{ days}$$

Recalling the following:

T_{sun} = (Number of Turns in the Sun)

$\#\, P_s$ = 13.35940210197 P_s

R_{mes} = 1(T_{sun}) (20 years)

R_{mes} = Y

If you consider only 13 without decimals you will has:

R_{mes} = (T_{sun}) (20 years)

R_{mes} = (13) (P_s) (20*365) days

R_{mes} = 13R_{me}

Replacing the values of each variables gets the exact 7000 years

$$Y = 13^{R_{me}}$$

Y = (13)(26.9230769230769230769230769230769230769231077*7300) days

Y = 2555000 days

The days convert into years:

Y = 2555000 (days) (1 year/365 days)

Y = 7000 years

V

Developing the formula with the Synodic period 29.530589 and considering leap year

$$W_{end} = ((\frac{(3n + l)P_s}{73(days)^2}) + S_d)$$
years

CENTURIA I: XLVIII

Explaining this centuria in a different way with the following formula:

If you consider Synodic period:

$$P_s = 29.530589 \text{ days}$$

$$R_{me} = (P_s)\ (20 \text{ years})$$

$$R_{me} = (P_s)\ ((15^n) + (5^l)) \text{ days}$$

$$R_{me} = (29.530589 * 7305) \text{ days}$$

$$X = 215720.952645 \text{ days}$$

$$X = 215720.952645 \text{ days } (1 \text{ year}/365 \text{ days})$$

Convert to years:

X = 591.01630861643835616438356164384 years

0.01630861643835616438356164383835
years (365 days/1 year)

= 5.9526449999999999999999999999775 days

Convert to hours:

0.9526449999999999999999999999775
days (24 hours/1 day)

= 22.863479999999999999999999999946 hours

Convert to minutes:

0.863479999999999999999999999946 hours
(60 minutes/1 hour)

= 51.80879999999999999999999999676 minutes

Convert to years:

0.808799999999999999999999676 ~~minutes~~
(60 seconds/1 ~~minute~~)

= 48.527999999999999999999998056 seconds

Approximately: X = 591 years, 5 days,   22
hours, 51 minutes, 48 seconds.

Recalling the following:

C_F = May 20

C_F = (Date of the Centuria IX - LXXXIII)

S_d = Subtraction of X and C_F

$S_d = X - C_F$

S_d = (1537 − 55) years, 5 months,   (14

ó 5) days, 1 hours, 8 minutes,   12

seconds

Recalling that:

$X = R_{me}$

We also know that:

$$R_{me} = 20P_s \text{ years} = \left(\frac{(3n + l)P_s}{73(days)}\right) \text{ years}$$

$$W_{end} = \left(\left(\frac{(3n + l)P_s}{73(days)}\right) \text{ years} + S_d\right)$$

Replacing the values of each variables gets the exact date of the end of the world.

$$W_{end} = (X + S_d)$$

Thursday, May 20, 2128

Tuesday, May 11, 2128

E

Developing the formula with
Synodic 29.530589 period and
without considering leap year

$$W_{end} = (\frac{(\frac{20P_s}{1day})}{} + S_d) \text{ years}$$

CENTURIA I: XLVIII

Explaining this centuria in a different way with the following formula:

$$R_{me} = (P_s)(20 \text{ years})$$

If you consider Synodic period:

$$P_s = 29.530589 \text{ days}$$

$$R_{me} = (P_s)(20 \text{ years})$$

$$R_{me} = (P_s)(20*365) \text{ days}$$

$$R_{me} = (29.530589*7300) \text{ days}$$

X = 215573.2997 days

X = 215573.2997 days (1 year/365 days)

Convert to years:

X = 590.61178 years

0.611779999999953 years (365 days/1 year)

= 223.299699999983 days

Convert to month:

223.299699999983 days (1 month/30 days)

= 7.44332333333276 months

Convert to days:

0.443323333332762 months (30 days/1 month)

= 13.299699999983 days

Convert to hours:

0.29969999998298 days (24 hours/1 day)

= 7.19279999959153 hours

Convert to minutes:

0.192799999591529 hours (60 minutes/ 1hour) = 11.5679999754917 minutes

Convert to seconds:

0.567999975491716 minutes (60 seconds /1 minute)

= 34.079998529503 seconds

Approximately: X = 590 years, 7 months, 13 days, 7 hours, 11 minutes, 34 seconds

Recalling the following:

$$C_F = \text{May 20}$$

$$C_F = \text{(Date of the Centuria IX - LXXXIII)}$$

$$S_d = \text{Subtraction of X and } C_F$$

$$S_d = X - C_F$$

$$S_d = \text{(1537 − 55) years, (10 ó 9) months, (6 ó 27) days, 16 hours, 48 minutes, 26 seconds}$$

Recalling that:

$$X = R_{me}$$

We also know that:

$$R_{me} = 20P_s \text{ years}$$

$$W_{end} = (20P_s \text{ years} + S_d)$$

Replacing the values of each variable gets the exact date of the end of the world.

$$W_{end} = (X + S_d)$$

Thursday, May 20, 2128

Tuesday, May 11, 2128

C

Developing the formula with the Synodic period 29.1666667 and considering leap year

$$W_{end} = \left(\left(\frac{(3n + l)P_s}{73(days)^2}\right) + S_d\right)$$
years

CENTURIA I: XLVIII

Explaining this centuria in a different way with the following formula:

$$R_{me} = (P_s)\ (20\ \text{years})$$

If you consider Synodic period:

$$P_s = 29.14670317134382842801733972166\ \text{days}$$

don't consider years leap

$$R_{me} = (P_s)\ (20\ \text{years})$$

$$R_{me} = (P_s)\ ((15^n) + (5^l))$$

$$R_{me} = \underline{(29.14670317134382842801733972166^*7305)}$$
$$\underline{\text{days}}$$

Convert to years:

X = <u>212916.66666666666666666667 days (1 year/365 days)</u>

X = 583.33333333333333333333333334 years

Converted years into days:

<u>0.33333333333333333333333334 years (365 days/1 year)</u>

= 121.666666666666666666666691 days

Convert to month:

<u>121.66666666666666666666691 days (1 month/30 days)</u>

= 4.0555555555555555555555556366 months

Convert to days:

0.0555555555555555555555555556366 ~~months~~ (30 days/1 ~~month~~)

= 1.6666666666666666666666669098 days

Convert to hours:

0.6666666666666666666666669098 ~~days~~ (24 hours/1 ~~day~~)

= 16.0000000000000000000000058352 hours

Convert to minutes:

0.0000000000000000000000058352 ~~hours~~ (60 minutes/1 ~~hour~~)

= 0.0000000000000000000003501120 minutes

Convert to seconds:

<u>0.00000000000000000000003501120</u>
<u>minutes (60 seconds/1 minute)</u>

= 0.0000000000000000000210067200 seconds

Approximately : X = 583 years, 4 months, 1
days, 16 hours, 0 minutes, 0 seconds

Recalling the following:

C_F = May 20

C_F = (Date of the Centuria IX - LXXXIII)

S_d = Subtraction of X and C_F

$S_d = X - C_F$

S_d = (1537 − 55) years, 1 months,   (18
ó 9) days, 7 hours, 59 minutes,   60
seconds

Recalling that:

$X = R_{me}$

We also know that:

$$R_{me} = 20P_s \text{ years}$$

$$R_{me} = \left(\frac{(3n + l)P_s}{\overline{73(days)}}\right) \text{ years}$$

$$W_{end} = \left(\left(\frac{(3n + l)P_s}{\overline{73(days)}}\right) \text{ years} + S_d\right)$$

Replacing the values of each variable gets the exact date of the end of the world.

$$W_{end} = (X + S_d)$$

Monday, May 20, 2120

Saturday, May 11, 2120

H

Developing the formula with Synodic 29.1666667 period and without considering leap year

$$W_{end} = (\frac{(\frac{20P_s}{1day})}{} + S_d) \text{ years}$$

CENTURIA I: XLVIII

Explaining this centuria in a different way with the following formula:

$$R_{me} = (P_s)\ (20\ \text{years})$$

If you consider Synodic period:

P_s = 29.16666666666666666666666666669 days
don't consider years leap

R_{me} = (29.16666666666666666666666666669*7300) days

X = 212916.66666666666666666666666 days (1 year/365 days)

Convert to years:

X = 583.33333333333333333333333333 years

0.33333333333333333333333333333 years (365 days/1 year)

= 121.66666666666666666666666665 days

Convert to month:

121.66666666666666666666666665 days (1 month/30 days)

= 4.05555555555555555555555555555 months

Convert to days:

0.05555555555555555555555555555 months (30 days/1 month)

= 1.66666666666666666666666665 days

Convert to hours:

0.66666666666666666666666665 days (24 hours/1 day)

= 15.99999999999999999999999996 hours

Convert to minutes:

0.99999999999999999999999996 hours (60 minutes/1 hour)

= 59.99999999999999999999999976 minutes

Convert to seconds:

0.99999999999999999999999976 minutes (60 seconds/1 minute)

= 59.99999999999999999999999856 seconds

Approximately: X = 583 years, 4 months, 1 days, 15 hours, 59 minutes, 59 seconds

Recalling the following:

C_F = May 20

C_F = (Date of the Centuria IX - LXXXIII)

S_d = Subtraction of X and C_F

S_d = X − C_F

S_d = (1537 − 55) years, 1 months,   (18 ó 9) days, 8 hours, 0 minutes,   1 seconds

Recalling that:

X = R_{me}

We also know that:

R_{me} = $20P_s$ years

$$W_{end} = (20P_s \text{ years} + S_d)$$

Replacing the values of each variable gets the exact date of the end of the world.

$$W_{end} = (X + S_d)$$

Monday, May 20, 2120

Saturday, May 11, 2120

A

CENTURIA I: XLVIII

Explaining this centuria in a different way with the following formula:

$$R_{mes} = 1(^{T}sun)\ (20\ \text{years})$$

$$R_{mes} = Y$$

Considering the value of the Synodic period:

$$P_s = 29.530589\ \text{days}$$

$$R_{mes} = (12.360065)\ (^{P}s)\ (20\ \text{years})$$

$$R_{mes} = (12.360065^{R}me)$$

$$Y = 12.360065^{R}me$$

$$Y = 2664499.9965564805\ \text{days}$$

The days convert into years:

Y = 2664499.9965564805 (days) (1 year/365 days)

Y = 7300 years

CENTURIA I: XLVIII

Explaining this centuria in a different way with the following formula:

$$R_{mes} = 1(^{T}sun)\ (20\ \text{years})$$

Considering the Synodic period:

$$P_s = 29.530589\ \text{days}$$

$$R_{mes} = Y$$

$$R_{mes} = (12.360065)\ (^{P}s)\ (20\ \text{years})$$

$$R_{mes} = (12.360065^{R}me)$$

$$Y = 12.360065^{R}me$$

$$Y = 2664499.99655648\ \text{days}$$

The days convert into years:

Y = 2664499.99655648 (days) (1 year/365 days)

Y = 7300 years

CENTURIA I: XLVIII

Explaining this centuria in a different way with the following formula:

$$R_{mes} = 1(T_{sun}) \ (20 \text{ years})$$

Recalling the following:

$$T_{sun} = (\text{number of turns in the Sun})$$

$$\# P_s = 12.360065 \ P_s$$

If you consider synodic period the value of P_s change to:

$$P_s = 29.16666666666666666666666666669 \text{days}$$

$$R_{mes} = 1(^T sun)\ (20\ \text{years})$$

$$R_{mes} = 1(^T sun)\ (20*365)\ \text{days}$$

$$R_{mes} = Y$$

Replacing the values of each variables gets the exact 7000 years :

$$R_{mes} = (12)\ ^{(P_s)}\underline{(7300)\ \text{days}}$$

$$R_{mes} = 12^{R_{me}}$$

Y = 2555000 days

The days convert into years:

Y = 2555000 (days) (1 year/365 days)

Y = 7000 years

N

CENTURIA I: XLVIII

Explaining this centuria in a different way with the following formula:

$$R_{mes} = 1(T_{sun}) ((15\ n) + (5\ l))$$

Considering the Synodic period:

$$P_s = 29.14670317134382842801733972166 \text{ days}$$

Recalling the following:

$$T_{sun} = (\text{Number of turns in the Sun})$$

$$\#\ P_s = 12.360065\ P_s$$

Replacing the values of each variables gets the exact 7000 years:

$$R_{mes} = 1(^{T}sun)(20 \text{ years})$$

$$R_{mes} = 1(^{T}sun)((15^{n}) + (5^{l}))\,^{R}mes$$

$$R_{mes} = Y$$

$$R_{mes} = (12)\,(^{P}s)\,\underline{(5475+1830)\ \text{days}}$$

$$R_{mes} = (12^{R}me)$$

Y = 2555000 days

The days convert into years:

Y = 2555000 (days) (1 year/365 days)

Y = 7000 years

The following right –to-left graph displays the sun and two dates for determine the exact day that the sun would take her tired days.

The first date is when the vision of the Centuria I – XLVIII it was revealed to Michelle.

The second dates are obtained with the formula Bárcena.

CENTURIA I : XLVIII

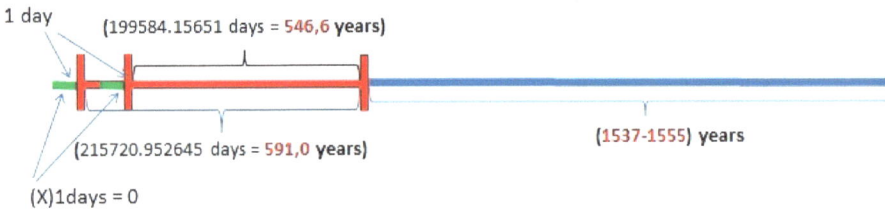

1 day

(199584.15651 days = 546,6 years)

(215720.952645 days = 591,0 years)

(X)1days = 0

(1537-1555) years

A-2

Figure A-2 shows from right to left the sum of two dates that determine the exact day that the Sun will take his weary days.

The first date is when the vision of the centuria was revealed to Michel de Nostradamus.

The second date is obtained with the formula Bárcena.

CENTURIA I : XLVIII

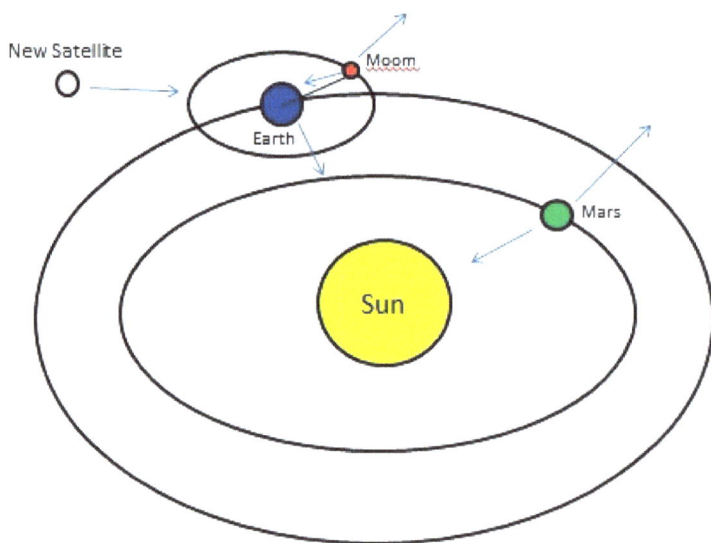

New Satellite
Moom
Earth
Mars
Sun

A-3

Explaining the table A-3 we show you completely around the Sun to the Earth with its natural satellite called the Moon.

In the future the moon will come out of its orbit and is replaced by another natural satellite.

The following formulas are described with the initials of the meaning in english and spanish respectively.

$$W_{end} = X + (^{S}d) \text{ years}$$

$$W_{end} = (\frac{R_{me}}{365(days)}) + ^{S}d) \text{ years}$$

$$W_{end} = (20^{P}s + ^{S}d) \text{ years}$$

$$W_{end} = ((\frac{(3n + l)P_{s}}{73(days)}) + ^{S}d) \text{ years}$$

$$M_{Fin} = X + (^{F}ini) \text{ años}$$

$$M_{Fin} = (\frac{R_{Le}}{365(dias)}) + ^{F}ini) \text{ años}$$

$$M_{Fin} = (20^{P}s + ^{F}ini) \text{ años}$$

$$M_{Fin} = ((\frac{(3n + b)P_{s}}{73(dias)}) + ^{F}ini) \text{ años}$$

$$M_{Fin} = (20P_s + F_{ini}) \text{ años}$$

$$W_{end} = (20P_s + S_d) \text{ years}$$

$$M_{Fin} = \left(\left(\frac{(3n + b)P_s}{73(dias)}\right) + F_{ini}\right) \text{ años}$$

$$W_{end} = \left(\left(\frac{(3n + l)P_s}{73(days)}\right) + S_d\right) \text{ years}$$

$$W_{end} = \left(\left(\frac{R_{me}}{365(days)}\right) + S_d\right) \text{ years}$$

$$M_{Fin} = \left(\left(\frac{R_{Le}}{365(dias)}\right) + F_{ini}\right) \text{ años}$$

$$W_{end} = X + (^S d) \text{ years}$$

$$W_{end} = (\dfrac{(\dfrac{R_{me}}{365day})}{} + ^S d) \text{ years}$$

$$W_{end} = ((20)P_s + ^S d) \text{ years}$$

$$W_{end} = ((\dfrac{(3n + l)P_s}{73days}) + ^S d) \text{ years}$$

$$M_{Fin} = X + (^F ini) \text{ años}$$

$$M_{Fin} = (\dfrac{(\dfrac{R_{Le}}{365dias})}{} + ^F ini) \text{ años}$$

$$M_{Fin} = ((20)P_s + ^F ini) \text{ años}$$

$$M_{Fin} = ((\dfrac{(3n + b)P_s}{73dias}) + ^F ini) \text{ años}$$

$$W_{end} = X + (^S d) \text{ years}$$

$$M_{Fin} = X + (^F ini) \text{ años}$$

G

CENTURIA IX

LXXXIII Day twenty of taurus so strong earth shaking,

The large crowded theatre will sink,

The air, land and sky darken and tremble

Then God with his saints to the unfaithful overwhelm

In this centuria, we get the day and month that helps us decrypt a part of the date of the world end and when Michelle wrote this centuria.

$W_{end} = X + (^S d)$ years

W_{end} = years number + 20 of Taurus

C_F = Month and day of the end = 20 of Taurus

If you add up the months, days, hours, minutes, seconds of $(X + ^S d)$ you will obtain the 20 of Taurus.

C_F = 20 of Taurus

$C_F = May\ 11$

$S_{d} = X - C_F$

Using the following formulas:

$$R_{mes} = 1(^T sun)(20 \text{ years})$$

$$W_{end} = (20P_s + S_d) \text{ years}$$

The first formula helps us to get the parameter that cannot be neither more nor less than (6669 < 7000 < 7001) years and with the second formula it will get a number of years and if you add the day and month of the Centuria I – LXXXIII can be determined when it was revealed his vision to Michel of Nostradamus and also know the date of the end of our days.

- X = 546 years, 9 months, 24 days,　　3 hours, 45 minutes, 14 seconds

- Sd = (1537-55) years, 7 months,　　(26 or 15)days, 20 hours, 14 minutes, 46 seconds

Saturday, May 20, 2084 or Thursday, May 11, 2084

- X = 546 years, 5 months, 7 days,　　13 hours, 9 minutes, 59 seconds

- Sd = (1537-55) years, (0 o 12) months, (12 or 3) days, 10 hours,　　50 minutes, 1 seconds

Thursday, May 20, 2083 or Tuesday, May 11, 2083

- X = 538 years, 5 months, 18 days, 11 hours, 4 minutes, 36 seconds

- Sd = (1537-55) years, (0 or 11) months, (1 or 22) days, 12 hours, 55 minutes, 24 seconds

  Monday, May 20, 2075 or Saturday, May 11, 2075

- X = 538 years, 5 months, 18 days, 11 hours, 4 minutes, 36 seconds

- Sd = (1537-55) years, (0 or 11) months, (1 or 22) days, 12 hours, 55 minutes, 24 seconds

  Monday, May 20, 2075 or Saturday, May 11, 2075

- X = 591 years, 0 month, 5 days, 22 hours, 51 minutes, 48 seconds

- Sd = (1537-55) years, 5 months, (14 or 5) days, 1 hour, 8 minutes, 12 seconds

Thursday, May 20, 2128 or Tuesday, May 11, 2128

- X = 590 years, 7 months, 13 days, 7 hours, 11 minutes, 34 seconds

- Sd = (1537-55) years, (10-9) months, (6 or 27) days, 16 hours, 48 minutes, 26 seconds

Thursday, May 20, 2128 or Tuesday, May 11, 2128

- X = 583 years, 4 months, 1 days,   16 hours, 0 minutes,   0 seconds

- Sd = (1537-55) years, 1 month,   (18 or 9) days, 7 hours, 59 minutes, 60 seconds

Monday, May 20, 2120 or Saturday, May 11, 2120

- X = 583 years, 4 months, 1 day,   15 hours, 59 minutes,   59 seconds

- Sd = (1537-55) years, 1 month,   (18 or 9) days, 8 hours,   0 minutes, 1 seconds

Monday, May 20, 2120 or Saturday, May 11, 2120

I have shown several dates of the end of the world but which one is true? For me it is the Monday, May 20, 2120 or Saturday, May 11, 2120 because they are described in the centuria exactly 20 years of reigns of the Moon that are 12 without considering the decimal and the constant 29.1666666666666666666666666666669 is the minor decrease decimal and without changing the whole numbers.

In 2013 we will find at the 18.3517% approximately when the Sun take their weary days.

The 18.3571% is approximately determined after getting the years with the following formula Bárcena

$$R_{me} = 20P_s \text{ years}$$

X = 100% of the time remaining to the end of the world.

If subtraction:

X - 18.39999999999999999999999999999% shall be equal to the number of years elapsed until 31 December 2012.

X = 583.3333333333333333333333333333 years
- 18.3999999999999999999999999999%

= 476 years.

The 476 years equals the 81.600000000....001%

(476<>81.600000000000000000000000000001%)

81.600000000000000000000000000001% is equal to the years up to the last seconds of 2012.

The other question about this constant 29.530589 is when will cease to be a constant and when will begin to decrease?

The answer is when that percentage becomes 1.232357179% more or less in the year 575.

This value is overridden in the formula Bárcena ($W_{end} = X + S_d$) and we have the year 2112.

After it convert decimals will have the month, day, hour, minutes and seconds approximately.

The percentage decreases the constant is equal to the number of years that the Sun will begin to lose energy and generate the planets slower roten or vice versa and the moon on the earth then its cycle slightly faster.

The decreased percentage of the constant is equal to the number of years that the Sun will start to lose power and generate the planets rotate slower or vice versa and do of the moon on the earth your cycle slightly faster.

The other alternative is when the remaining percentage until the end of the world is 0,6% this turn into years and we have obtained the following date:

579 years, 10 months, 1 days, 15 hours, 59 minutes, 59 seconds.

Then if we replace this value in the formula Bárcena we will obtain the following date:

2116 years, 11 months, (20 or 11) days, 0 hours, 0 minutes, 0 seconds.

I take as reference this 0,6% not elapsed for the end of the world because in the chapter 6, 13, 11, 21, 22 of the revelation of John in the New Testament Bible. There is described the 42 months that the beast will have to do all that is him to allowed to deceive men on earth and put them against God after that time God and his legions of angels to do will overwhelm to anyone that is in the service of the beast.

When that percentage came to be equal to 0% the new constant value will be equal to 29.1666666666666666666666666666669 then the prophecy will be consumed.

$$W_{end} = X + S_d$$

- X = 583 years, 4 months, 1 day, 15 hours, 59 minutes, 59 seconds

- S_d = (1537-55) years, 1 months, (18 ó 9) days, 8 hours, 0 minutes, 1 seconds

  Monday, May 20, 2120

  Saturday, May 11, 2120

123

E

If you consider mes = 30.41666666666666666666 666666665 days we will obtain the following dates.

- X = 546 years, 9 months, 20 days,   9 hours, 45 minutes, 14 seconds

- Sd = (1537-55) years, 7 months,   (30 or 19)days, 14 hours, 14 minutes, 46 seconds

Saturday, May 20, 2084 or
Thursday, May 11, 2084

- X = 546 years, 5 months, 3 days,   19 hours, 9 minutes, 59 seconds

- Sd = (1537-55) years, (0 o 12) months, (16 or 7) days, 4 hours,   50 minutes, 1 seconds

Thursday, May 20, 2083 or
Tuesday, May 11, 2083

- X = 538 years, 5 months, 14 days, 17 hours, 4 minutes, 36 seconds

- Sd = (1537-55) years, (0 or 11) months,   (5 or 27) days, 6 hours, 55 minutes, 24 seconds

  Monday, May 20, 2075 or Saturday, May 11, 2075

- X = 538 years, 5 months, 14 days,   17 hours, 4 minutes, 36 seconds

- Sd = (1537-55) years, (0 or 11) months, (5 or 26) days,   6 hours, 55 minutes, 24 seconds

  Monday, May 20, 2075 or Saturday, May 11, 2075

- X = 591 years, 0 month, 2 days, 4 hours, 51 minutes, 48 seconds

- Sd = (1537-55) years, 5 months, (17 or 8) days, 1 hour, 8 minutes, 12 seconds

  Thursday, May 20, 2128 or
  Tuesday, May 11, 2128

- X = 590 years, 7 months, 9 days, 13 hours, 11 minutes, 34 seconds

- Sd = (1537-55) years, (10-10) months, (10 or 1) days, 10 hours, 48 minutes, 26 seconds

  Thursday, May 20, 2128 or
  Tuesday, May 11, 2128

- X = 583 years, 4 months, 27 days, 22 hours, 0 minutes,   0 seconds

- Sd = (1537-55) years, 2 month,   (22 or 13) days, 1 hours, 59 minutes, 60 seconds

  Monday, May 20, 2120 or Saturday, May 11, 2120

- X = 583 years, 3 months, 27 day, 21 hours, 59 minutes,   59 seconds

- Sd = (1537-55) years, 2 month,   (22 or 13) days, 2 hours,   0 minutes, 1 seconds

  Monday, May 20, 2120 or Saturday, May 11, 2120

Before finishing this book I want to tell you that the initials that we have throughout these pages are two words that will help you to be inside the good harvest in recent days.

The first word is the love to God and to all good people surrounding him and bad people aevil carefully and pidele God that you free everything bad.

In addition to love yourself, this means improve yourself on all levels of your life and also emotionally, physically, humanitarian and remember that everything is to do good must be within the divine justice.

The second word is change in your life if these bad way, if you don't want to know

God, yet have little time to make things right if you have faith in God.

Read the Bible there this the guide and the rules that God command for that to be present in your harvest.

No tattooing the name or number of the beast never in your body.

In the Bible in the revelation of John you explains why not serve nor worship the beast.

It passes this message from generation to generation.

That this God with you from generation in generation and by the centuries of centuries.

Amen

The future it is wonderful not feel panic, the final judgment is very close after waiting millions of years the dead may have eternal life if this within the harvest of God.

Monsters of soul never will be with us again and the wonderful thing is that God will come to dwell in the land forever.

We can travel the whole universe in the future with God.

THE END

$$R_{mes} = 1(T_{sun})(20 \text{ years})$$

$$R_{mes} = 1(T_{sun})((15\,n) + (5\,l))$$

$$R_{mes} = 12.360065\ R_{me}$$

$$R_{mes} = Y * 1\text{day}$$

$$R_{Les} = 1(V_{Sol})(20 \text{ años})$$

$$R_{Les} = 1(V_{Sol})((15\,n) + (5\,b))$$

$$R_{Les} = 12.360065\ R_{Le}$$

$$R_{Les} = Y * 1\text{días}$$

$$R_{mes} = 1(T_{sun})(20 \text{ years})$$

$$R_{Les} = 1(V_{Sol})(20 \text{ años})$$

$$R_{mes} = 1(T_{sun})((15\,n) + (5\,l))$$

$$R_{Les} = 1(V_{Sol})((15\ n) + (5\ b))$$

$$R_{mes} = 12.360065\ R_{me}$$

$$R_{Les} = 12.360065\ R_{Le}$$